WHY SHOULD I EAT THIS?

WHOLE GRAINS

TABLE OF CONTENTS

By Lindsay Snea

A Crabtree Seedlings Book

Crabtree Publishing

crabtreebooks.com

School-to-Home Support for Caregivers and Teachers

This book helps children grow by letting them practice reading. Here are a few guiding questions to help the reader with building his or her comprehension skills. Possible answers appear here in red.

Before Reading:

- What do I think this book is about?
  - *I think this book is about food.*
  - *I think this book is about whole grains.*
- What do I want to learn about whole grains?
  - *I want to learn what kinds of plants give us whole grains.*
  - *I want to learn what kinds of foods contain whole grains.*

During Reading:

- I wonder why...
  - *I wonder how whole grains give our bodies energy.*
    I wonder why some foods are made with refined grains.
- What have I learned so far?
  - *I have learned that whole grains contain a nutrient called fiber.*
  - *I have learned that whole grains are good for our hearts and brains.*

After Reading:

- What details did I learn about whole grains?
  - *I have learned that foods with refined grains are not as healthy as food with whole grains.*
  - *I have learned that we should eat whole grains every day.*
- Read the book again and look for the vocabulary words.
  - *I see the word **vitamins** on page 4, and the word **refined** on page 14. The other glossary words are found on pages 22 and 23.*

WHOLE GRAINS

Grains are the dried seeds of grass-like plants. Grain plants include wheat, barley, corn, and rice.

Foods that are whole grain contain all the parts of a plant's seed. The seeds give you important **vitamins** and **nutrients** your body needs to be healthy.

Whole grains contain an important nutrient called fiber. It helps to move food through your body.

Whole grains are good for your heart and your brain. They also help your bones grow strong.

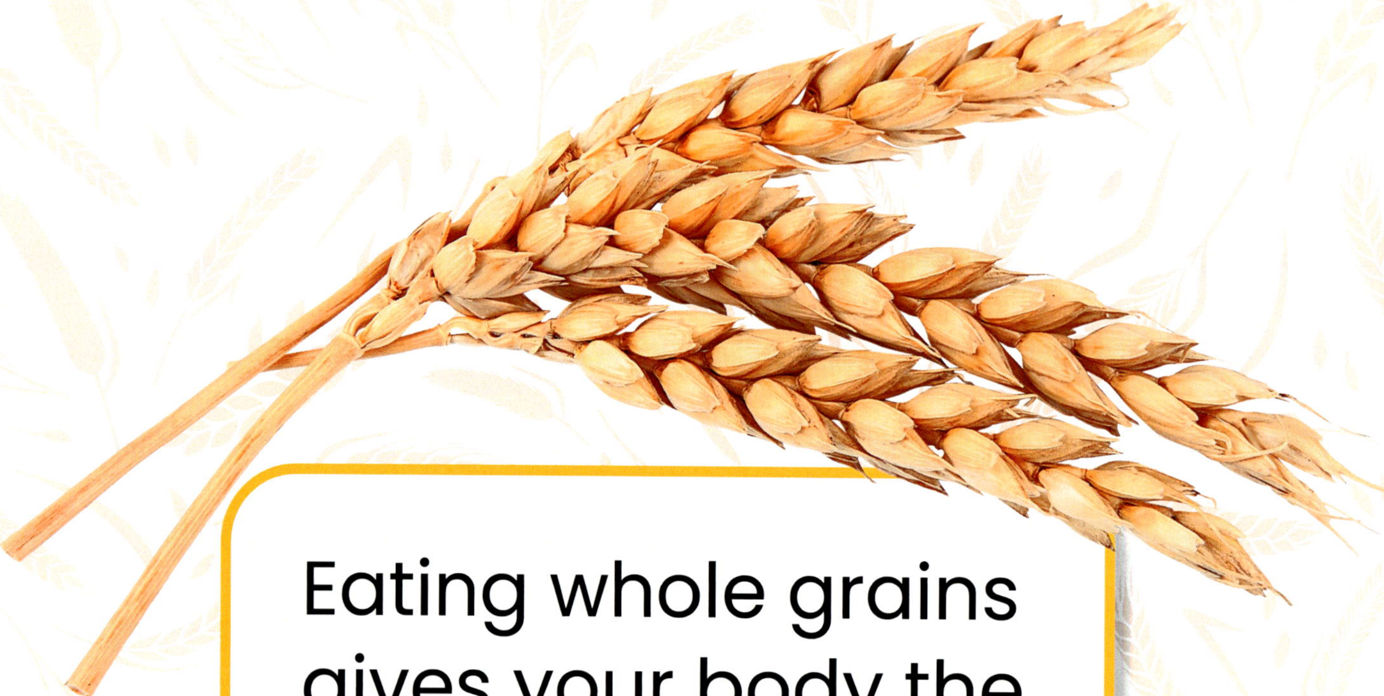

Eating whole grains gives your body the **energy** it needs to play, to learn, and to grow.

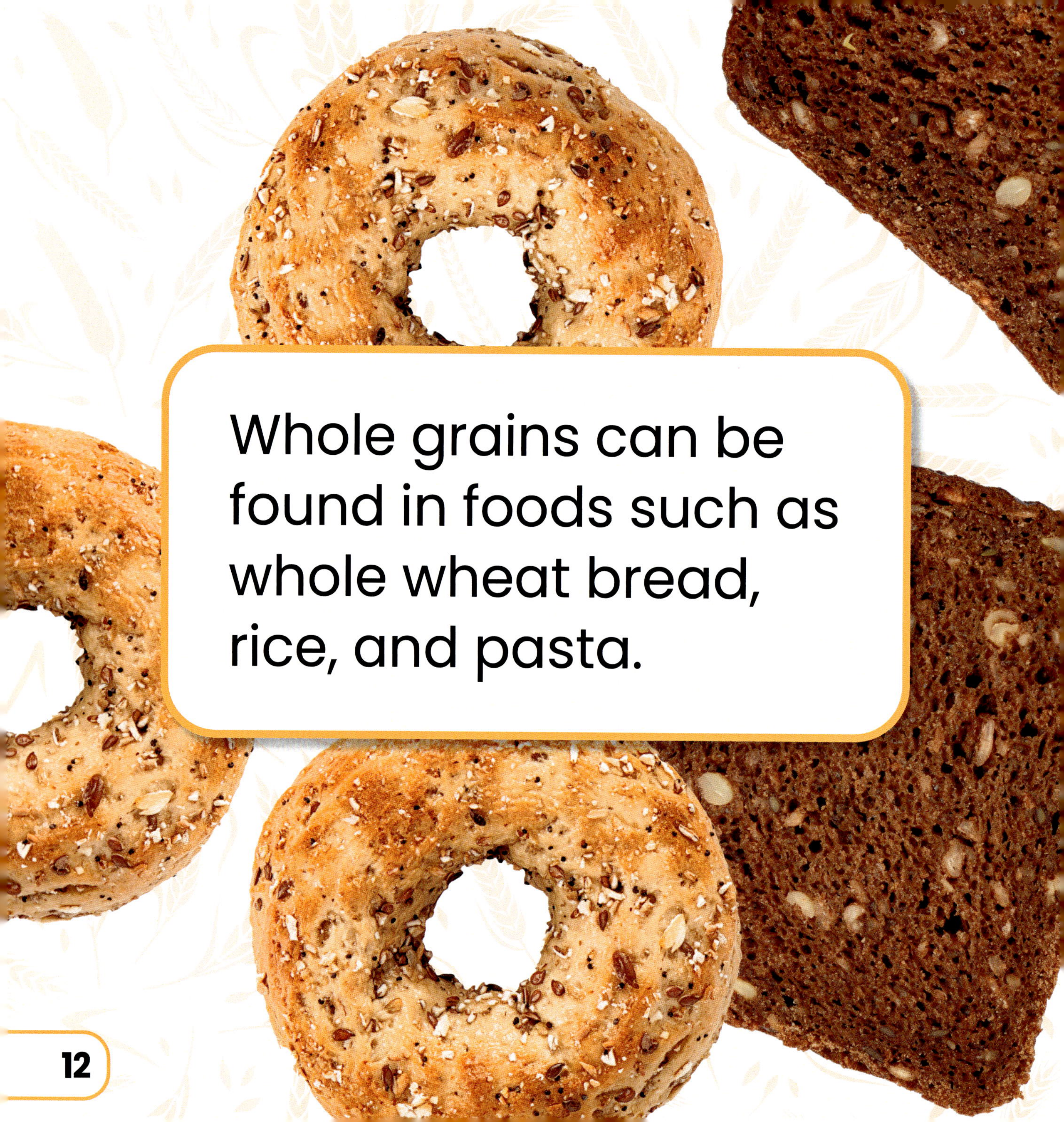

Whole grains can be found in foods such as whole wheat bread, rice, and pasta.

Some foods contain **refined** grains. These foods are made with only part of the seed. Refined grains do not have as many healthy nutrients as whole grains.

DID YOU KNOW?

Foods made from refined grains include white bread, white rice, pastries, and crackers.

You should eat whole grains every day. Try to make them a part of most meals and snacks.

Meals made with whole grains could include oatmeal for breakfast, a sandwich made with whole grain bread for lunch, and whole wheat pasta for dinner.

DID YOU KNOW?

Popcorn is a healthy whole grain snack.

Whole grains are an important part of a healthy **diet**.

Glossary

diet (DAHY-it): The food that you usually eat

energy (EN-er-jee): Your body or mind's ability to be active or do work

nutrients (NOO-tree-uhnts): Substances that your body needs to live and grow

refined (ri-FAHYND): Refers to grains that have had parts of the seeds removed to make them softer and last longer

vitamins (VAHY-tuh-mins): Nutrients found in foods that help your body grow and stay healthy

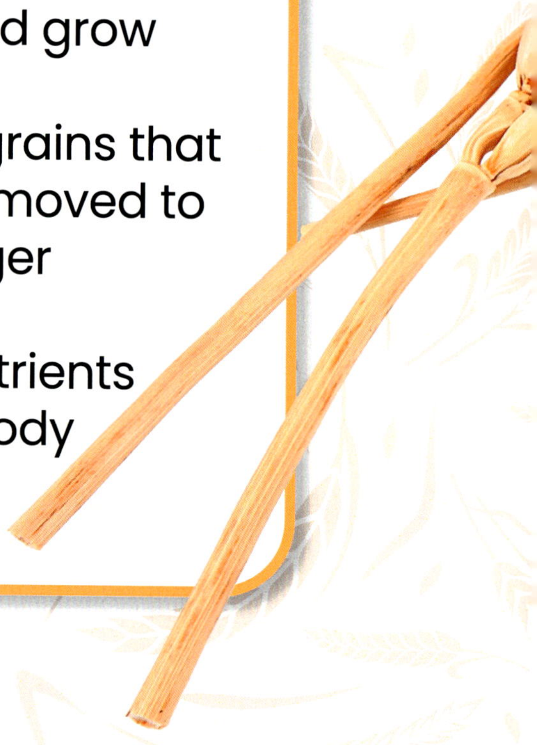

Index

About the Author

Lindsay Snea lives in the mountains of Fernie, British Columbia, Canada, with her husband and four children. Lindsay is an elementary school teacher, and a lifelong learner, who is passionate about cooking, food, fitness, being outdoors, and health.

Written by: Lindsay Snea
Designed by: Margaret Salter
Proofreader: Melissa Boyce
Photographs: All images from Shutterstock
Production manager: Candice Campbell

Crabtree Publishing

crabtreebooks.com 800-387-7650

 In Canada: We acknowledge the financial support of the Government of Canada through the Canada Book Fund for our publishing activities.

Printed in the U.S.A./CP112025

Published in Canada
Crabtree Publishing
616 Welland Avenue
St. Catharines, Ontario
L2M 5V6

Published in the United States
Crabtree Publishing
347 Fifth Avenue
Suite 1402-145
New York, New York, 10016

Library and Archives Canada Cataloguing in Publication
Available at Library and Archives Canada

Library of Congress Cataloging-in-Publication Data
Available at the Library of Congress

Hardcover: 978-1-0398-8118-1
Paperback: 978-1-0398-8479-3
Ebook (pdf): 978-1-0398-8238-6
Epub: 978-1-0398-8358-1
Read-Along: 978-1-0398-8563-9
Audio: 978-1-0398-8521-9